HOW I FOUGHT AGAINST MY PANCREATIC TUMOR

DR. JIAJIU SHAW

ILLUSTRATIONS BY CINDY Y. SHAW

DISCLAIMER

The contents provided in this book are presented to the best of my memory and are for information and reference purposes only. This book does not constitute the practice of any medical, nursing, or other medical professional health care advice, diagnosis, or treatment; nothing in the book should be used as a substitute for medical advice, diagnosis, or treatment.

Although some commercial products are mentioned in this book, it does not mean that the author endorses the specific products. Furthermore, none of the contents in this book represents or warrants that any particular product or service is appropriate or effective for any diseases.

Readers shall always seek the advices of their physicians or other medical practitioners with any questions regarding personal health or medical conditions and be responsible for their own health decisions made and/or activities taken. The author expressly disclaims all liability for damages of any kind arising out of reliance on any information of this book.

HOW I FOUGHT AGAINST MY PANCREATIC TUMOR

Published by Shaw Creative, Inc.
Henderson, Nevada

Illustrations by Cindy Shaw

ISBN 978-0-9904226-5-5

Printed in the United States of America

Preface

This book is related to my encounter with a specific pancreatic tumor, Intraductal Papillary Mucinous Neoplasm (IPMN). After my pancreatic IPMN was serendipitously discovered, I immediately considered taking drastic actions. One year later, I was lucky to be declared IPMN free. Since my experience is not a controlled study, it should not be treated as a rigorous scientific study but rather as an individual's real-life experience.

My goal in writing this book is to let readers know my story firsthand about the details of my unexpected journey and what I did that resulted in my success. Hopefully readers can follow my story and enjoy it.

I also hope this book can help readers understand more about pancreatic IPMN and the potential of natural products. However, it is very important for readers to know that this book is by no means intended to substitute the expertise of readers' physicians or other medical practitioners. For any medical decision, readers should consult with their physicians or other medical specialists.

In addition, I would like to show my appreciation to all my collaborators, especially Dr. Frederick Valeriote, Dr. Ben Chen, and Dr. Stephen Brown; all of them have been instrumental in my cancer research career. Comments from Dr. John Anagli, Dr. Stephen Brown, Dr. Ben Chen, Mrs. Chingchia Leong, Mr. Ian Leong, Mr. Charles Liu, Mr. Joseph Media, and Dr. Frederick

Valeriote are also very much appreciated. Furthermore, many thanks to my talented daughter, Cindy, for her work in making the illustrations and editing the manuscript.

Finally, if you enjoy this book in any way, please share it with your loved ones and friends.

Sincerely,

Jiajiu Shaw, Ph.D.

HOW I FOUGHT AGAINST MY PANCREATIC TUMOR

Table of Contents

Chapter 1

Beginning of the Story

October 24, 2015 was a very significant day in my life. I underwent an MRI (Magnetic Resonance Imaging) to understand the origin of an abdominal problem. Within a couple of days, the test results (Fig. 1.1) were available. Unexpectedly, the results showed a 0.5 cm tumor in the neck of my pancreas. This pancreas tumor was called a side-branch Intraductal Papillary Mucinous Neoplasm (IPMN). The official clinical impression was:

> *A 0.5 cm cystic lesion in the pancreatic neck, likely a sidebranch IPMN. This can be followed up in one year to determine stability.*

Further narrative of the findings is quoted below:

> *There is a 0.5 cm cystic lesion in the pancreatic neck (series 401 image 22), most likely side branch intraductal papillary mucinous neoplasm. No main pancreatic ductal dilatation.*

The discovery of my IPMN came as a total surprise to me. Immediately after reading the test results, I felt shocked and mentally at a loss. Although I was over sixty years old, I had been living a pretty healthy life and never thought I'd face a diagnosis of pancreatic tumor someday.

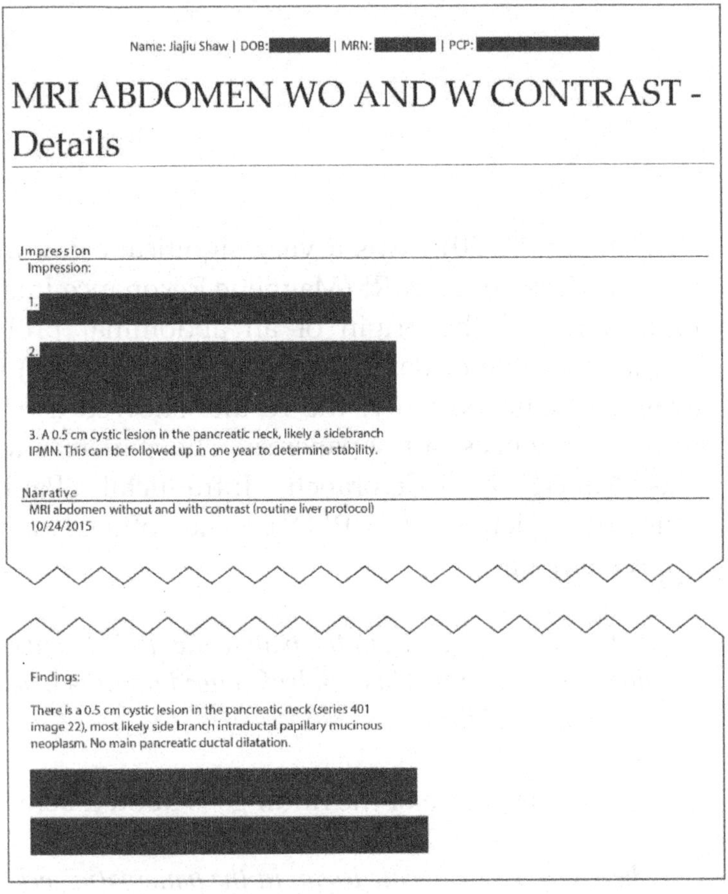
Name: Jiajiu Shaw | DOB: [redacted] | MRN: [redacted] | PCP: [redacted]

MRI ABDOMEN WO AND W CONTRAST - Details

Impression

Impression:

1. [redacted]

2. [redacted]

3. A 0.5 cm cystic lesion in the pancreatic neck, likely a sidebranch IPMN. This can be followed up in one year to determine stability.

Narrative

MRI abdomen without and with contrast (routine liver protocol)
10/24/2015

Findings:

There is a 0.5 cm cystic lesion in the pancreatic neck (series 401 image 22), most likely side branch intraductal papillary mucinous neoplasm. No main pancreatic ductal dilatation.

[redacted]

Fig. 1.1 Results of my MRI on 10/24/2015

Nevertheless, the discovery of this IPMN answered my 2-year old question on why my lower abdomen had been bothering me. For more than two years, I had been suffering subtle and sporadic dull abdominal pain. The pain was usually at the lower abdomen but did not always appear at the same location. Although the abdominal pain was not severe, it gradually became more frequent and was really bothering me.

Not knowing why it was happening and what to do, I had some fear that something bad might be happening. Another warning sign was that I had lost about 5 lb from 2014 to 2015, which could be another sign of some serious health issue, or maybe just anxiety.

To look into this issue, my primary physician ordered an X-ray test for me for two consecutive years. However, nothing unusual was identified in my abdomen by the X-ray tests. Therefore, my "mysterious abdominal pain" remained an unanswered question.

In 2015, I changed my health insurance company and also changed my primary physician. After I had a short discussion with my new physician, he recommended that I should have an ultrasound (sonogram) to examine my internal organs including liver and pancreas. Although there were some cysts identified on my liver, they did not appear to be a serious problem. As for my pancreas, the ultrasound results were not very clear and did not answer my question, thus my physician ordered an MRI for me. October 24, 2015 was the day my first MRI was conducted.

This MRI test again revealed some issues related to my liver (several cysts), similar to those indicated by ultrasound, but that was not my main concern. The most concerning to me was my pancreatic abnormality. I believed that this MRI might have revealed the answer I had been looking for.

My MRI results disclosed a critical health issue, pancreatic IPMN. If left untreated, there was a good chance that my IPMN could progress to become an invasive pancreatic cancer. If this were to happen, it could be the end of the world for me.

Like most people, I had never thought that this could happen to me, a tumor on my pancreas. Right away, I was confused and many questions came to my mind. How much longer would I be living? What should I do? What could I do?

Because of my concern, my physician recommended that I should see a gastrointestinal (GI) specialist for consultation. As a result, I met with a GI specialist and had a short discussion with him. His opinion was that the tumor was still small and I could just wait and monitor it annually. However, deep in my mind, I was very worried because, no matter the size, pancreatic IPMN could be very dangerous; therefore, I didn't want to just wait and do nothing.

Out of desperation, I decided to do some research myself and to take some actions to fight against this IPMN. My goal, although naïve, was to shrink my pancreatic tumor significantly in one year. Meanwhile, I also decided to write down my actions so that my experience might be useful for myself and people who might be in a similar situation.

I wasn't sure what was going to happen to me with my pancreatic tumor. However, I did believe (and I had to) that I should have at least one year to live. Therefore, I decided to do several important things in the following year:

(1) to do some research immediately and have a better understanding about pancreatic IPMN,
(2) to take some drastic therapeutic actions and hope to see the shrinkage of my pancreatic tumor, and
(3) to write down my activities against my pancreatic tumor.

After doing some research and planning, I started my desperate plan; I also started to write down my activities and health condition starting from January 8, 2016. Basically, that's pretty much what this book is all about.

Chapter 2

My Background and Training

Before we go further into my activities and my journey in dealing with my IPMN, let me give you a brief introduction about me and my background as a point of reference to my story.

My Background

When I was in high school, cancer struck my father in his throat. I believed it was related to his smoking habit because he was a heavy smoker; smoking was a popular social activity in the 1960s.

This was my first encounter with cancer and it made a lasting impression on me. As a teenager, I was deeply saddened and this incident impacted me greatly. Even today, half a century later, I still remember the moment that my mother took my father to a hospital to be treated by radiotherapy. The depressing mood around my whole family is a vivid memory. As a result, I'm sure this shaped my life decisions and my deep convictions against this dreaded disease.

Not too long after I graduated from the Chemistry Department at National Tsing Hua University in

Taiwan, one of my classmates died of cancer. From that time on, some of my relatives or acquaintances either died of or suffered from cancer, including my mother-in-law who died of uterus cancer. In 2007, I suffered another heart-breaking loss, my mother died of liver cancer. All of these happenings made my desire to work in cancer research stronger even though I knew my impact would be very limited.

After I graduated from Tsing Hua University in 1972, I did my compulsory military service and then worked as a high school teacher for a short time. Subsequently, I worked in the polymer industry in the development of coating, adhesive, and ink. For several years, I really did not have any chance to get involved in cancer research. However, some of my experience in these years (e.g., formulation of coating and ink) was somewhat helpful in my drug formulation development activities. Nevertheless, I always kept in mind that someday I would like to be involved in cancer research.

My Experience and Training

Seven years after I received my B.S. degree, I got tired of doing the work that was not challenging. In the meantime, some of my classmates received their Ph.D. degrees and were already involved in basic and applied research work. I figured it might be too late if I didn't change my career soon. As a result, I decided to go back to school to start my graduate work. In 1979, I became a student again at Texas A&I University (now A&M) in Analytical Chemistry. With a graduate assistantship granted to me, I transferred to the University of Kansas, Lawrence, Kansas (KU) one year later.

As luck would have it, one of the research projects assigned by my mentor and advisor, Dr. Grover Everett Jr., at KU was to synthesize analogs of cisplatin, which was a popular anticancer drug during the 1980s. As part of my graduate research, I was able to make several new analogs of cisplatin successfully. Eventually, my research work on this particular project became the beginning of my new drug research career.

After receiving my Ph.D. degree from KU, I did my postdoctoral research at the University of North Carolina, Chapel Hill (UNC), where I learned more about synthetic skills in organic and in organometallic chemistry. Again, this became part of the basic training that strengthened my ability in cancer drug design and synthesis.

Subsequently, I worked at several small and large pharmaceutical companies in Maryland, New Jersey, Pennsylvania, and Michigan, including Ciba-Geigy Corporation (now Novartis). Over these years, I learned more about analytical chemistry, formulation, and other preclinical development studies related to drug development. My experience in the pharmaceutical industry in this period of my life laid down the foundation of my venture in new drug discovery and development.

In 1997, I founded Unitech Pharmaceutical Inc. to focus on the discovery and development of anticancer agents. During the time between 1997 and 2007, I synthesized a series of anticancer agents, received several SBIR grants from the National Institutes of Health (NIH), and more than ten patents in the area of anticancer agents. In 2008, I joined 21st Century Therapeutics Inc. and continued my research endeavor in cancer and immune-mediated diseases.

To date, I have received over $3 million of research grants from the NIH and other government agencies. I also published a number of papers in peer-reviewed scientific journals and received 18 patents, mostly related to cancer and related areas.

In 2015, I wrote a book, Is Cancer in Us, to help the general public understand more about cancer. In 2016, I co-authored another book, Side Effects of Chemotherapy and Radiotherapy, with Dr. Frederick Valeriote and Dr. Ben Chen. This book was written because I couldn't find a book with a complete coverage on the subject of side effects of chemotherapy and radiotherapy.

With my background, training, and experience, I believed I could come up with a plan to counter my pancreatic tumor, albeit there was only a small chance that my plan could work.

Chapter 3

What about My Pancreatic Tumor, IPMN

In order to learn more about pancreatic IPMN, it is a good idea to first know some background information related to the pancreas.

What about the Pancreas?

The pancreas is a long flattened organ (about six to ten inched long) located in the upper left abdomen. It is located behind the stomach, and surrounded by liver, intestine, spleen, and gallbladder (Fig. 3.1).

The pancreas is in a shape similar to a fish lying sideways in the upper abdomen. The head of the pancreas (fish head) is positioned near the juncture of the stomach and the small intestine. The center part is called the body or the neck of the pancreas (fish neck). The small part extends to the left is called the tail of the pancreas (fish tail).

(a)

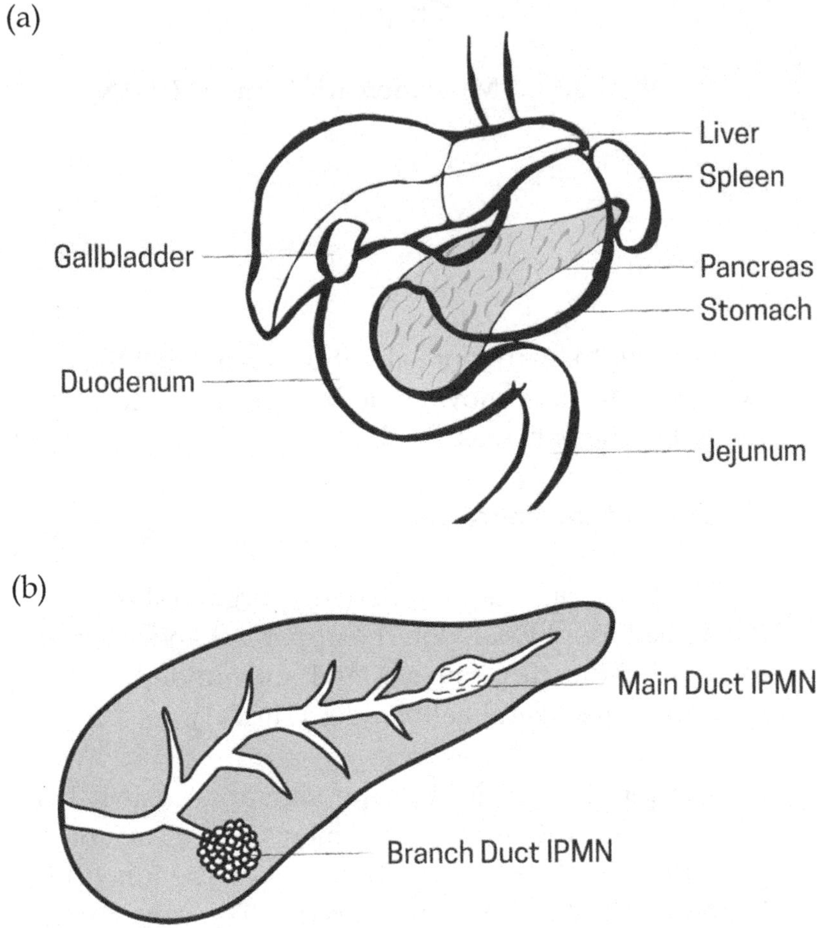

Fig 3.1 Schematic representations of (a) the pancreas and its location, (b) examples of a main-duct IPMN and a branch-duct IPMN

Functions of the Pancreas

The functions of the pancreas can be divided into two general categories shown below:

Exocrine Function - external digestive role

Most of the pancreas (about 95%) consists of exocrine cells that produce important enzymes for digestion. These enzymes include trypsin, chymotrypsin, amylase, and lipase for the digestion of protein, carbohydrate, and fat. These enzymes are released into branch duct, moved through main duct, and eventually into the duodenum, the first part of the intestines.

Endocrine Function - internal hormone role

The endocrine pancreas consists of small islet cells; islets are clusters of cells, with each "islet" containing 3,000 to 4,000 cells. It is estimated that there are 1 million islets in an adult pancreas. Islet cells release important hormones (instead of enzymes) into the bloodstream. These hormones include insulin and glucagon; both are critical in maintaining proper levels of blood sugar.

In summary, the pancreas is a very important organ for both the digestive system and blood sugar control. Theoretically, certain digestion issues or blood sugar control could be logically related to the pancreas. For example, chronic pancreatitis (inflammation of pancreas) impairs the body's ability to digest food and/or regulate blood sugar.

What about Pancreatic IPMN?

With a basic understanding of the pancreas and its functions, we are ready to have a general understanding about pancreatic IPMN. What is pancreatic IPMN? As mentioned before, IPMN is the abbreviation of Intraductal Papillary Mucinous Neoplasm. Let's give IPMN a clearer description here.

What is a papilla? Papilla means a nipple-like structure. How about mucin? Mucins are gel-like secretions comprising high molecular proteins. How about neoplasm? Neoplasm means new and uncontrolled cell growth (different from that of normal cells) that is not under normal physiologic control. Neoplasms can be either benign or malignant.

What does it mean by intraductal? It takes a few more sentences to explain this.

As shown in Fig. 3.1(b), for most people, there is one main duct on the pancreas that joins the pancreas to the common bile duct to supply pancreatic juice provided by the exocrine pancreatic tissue. In addition, there are a few branch ducts stretching outwards from the main duct, similar to a few branches stretching outwards from a tree trunk.

IPMNs are one group of particular tumors that grow within the pancreatic ducts; thus they are called intraductal neoplasms. They are a group of intraductal mucin-producing cystic neoplasms of the pancreas with clear malignant potential. If left untreated, IPMN can become malignant pancreatic cancer.

The World Health Organization (WHO) defines IPMNs as intraductal mucin-producing neoplasms with tall, columnar, mucin-containing epithelium with or without papillary projections.

IMPNs can be divided into main-duct IPMN, branch-duct IPMN, or both. Patients with branch-duct lesions are at lower risk for developing malignancy; patients with IPMNs involving the main duct are at a higher risk [1]. A schematic representation of main-duct IPMN and branch-duct IPMN is shown in Fig. 3.1(b).

Here are two interesting facts about pancreatic IPMN:

(1) IPMNs in pancreas are similar to polyps in colon. When polyps are discovered in colon during a colonoscopy, they are normally removed during the test. This is because it is well known that polyps can develop into colon cancer. Unfortunately, IPMNs are not as easily removed as polyps.

(2) Similar to colon cancer that occurs more often in older people, IPMN occurs more often for people over the age of 60s, and it happens more commonly in men [2].

What Are the Symptoms of IPMN?

Because of the deep location of the pancreas, tumors of the pancreas are not palpable. This explains why most pancreatic tumors are not detected until the tumors have grown large enough to be felt by pressing on the abdomen.

The other unfortunate fact is that symptoms of pancreatic cancer are not clear until it is too late. The most common symptoms of severe IPMN are listed below [3, 4]:

Pancreatitis, abdominal pain, nausea, vomiting, yellow skin or eyes (from the buildup of bilirubin, a waste product), fatigue, weight loss, fever, night sweats.

Once a patient is suspected to have an IPMN, the suspicion can be confirmed by either computerized tomography (CT) or MRI. Sometimes, a fine needle aspiration (FNA) biopsy can be obtained to confirm the diagnosis.

Pancreatic IPMN is not easily diagnosed and could become pancreatic cancer if left untreated [4]. Also, pancreatic IPMN could come back and become pancreatic cancer even after it is resolved. The unfortunate fact is that by the time a patient is diagnosed with pancreatic cancer, the disease is often already in an advanced stage.

Known Methods for Treating IPMN

IPMNs of the main pancreatic duct are usually treated by surgical excision. The following quote from Khan et al. concisely describes the general practice for the treatment of IPMN [3]:

Surgical resection is the recommend treatment for patients with high-risk features while close observation can be offered to patients without worrisome signs and symptoms of carcinoma.

According to the paper published by Sohn et al., in 2004, for main-duct IPMN, the treatment of choice is resection because it has about 50% chance of becoming malignant. On the other hand, side branch IPMNs can be monitored with regular CT or MRI, but most are eventually resected, with a 30% chance of malignancy in these resected tumors [5]. The following is quoted from this paper:

Between January 1987 and March 2003, inclusive, 136 pancreatic resections were performed for patients with IPMNs, with 78 resections performed since January 2001. ... IPMNs without evidence of invasive cancer were identified in 62% (n = 84) of patients... The remaining 38% (n = 52) of patients had IPMNs with associated invasive cancer..... The overall 5-year survival for patients having IPMNs without invasive cancer was 77%..... For those patients with invasive IPMNs, 2-year survival was 40% when margins were positive for invasive cancer or for IPMN without invasive cancer, and 60% when margins were tumor-free (P = *0.15).*

Similar to the above publication, according to Jang et al., most clinicians agree that surgical resection is required to treat main-duct IPMNs because of its high malignancy rate [6]. In Korea, for branch-duct IPMNs, if they are malignant, surgical treatment is recommended. However, for branch-duct IPMNs, if the tumor size is ≤ 2 cm, observation is recommended.

Pancreatic surgery can include the removal of the head of the pancreas, removal of the body and tail of the pancreas, or rarely removal of the entire pancreas. In some cases the surgery can be performed using minimally invasive techniques such as laparoscopy or robotic surgery [7].

A recent report showed that a patient with main duct IPMN was treated with photodynamic therapy (PDT). Porfimer sodium (a sensitizing agent activated by light used in the treatment of tumors) was administered intravenously, and laser light was delivered by a diffusing catheter placed in the pancreatic duct. Results from imaging and biopsy indicated that the IPMN was resolved and symptoms were also resolved, although metastatic cancer was diagnosed 2 years after PDT had been initiated. This study, however, indicated that pancreatic PDT was well tolerated in this case, and may be a therapeutic option for selected patients with main duct IPMN [8].

Known Drugs Effective at Treating IPMN?

At the present time, there is no drug approved by the FDA for the treatment of IPMN. However, there are a number of drugs approved by the FDA for treating pancreatic cancer. These include everolimus, erlotinib, gemcitabine, mitomycin C, and sunitinib.

In addition, there are also drug combinations approved for the treatment of pancreatic cancer, including the following:

FOLFIRINOX: FOL (folinic acid), F (fluorouracil), IRIN (irinotecan), OX (oxaliplatin)
Gemcitabine and Cisplatin
Gemcitabine and Oxaliplatin

Pancreatic cancer is a devastating cancer. Not only has it the lowest survival rate among all cancers, but also there is no good drug to eradicate the cancer although there are a number of drugs approved for pancreatic cancer. Therefore, no one wants to hear a diagnosis of pancreatic cancer. In this regard, IPMN is a critical pancreatic issue because, for many patients, IPMN eventually becomes pancreatic cancer, especially the main duct IPMN [4, 9].

Chapter 4

What Did I Do?

My Special Plan

Although my GI specialist told me not to worry about my IPMN because it was still small, I myself was very worried. Since I was told to just monitor it annually without any special actions, I worried even more. However, simply worrying about it was not going to solve this problem, which was very concerning to me. Therefore, I was determined to do something drastic to combat my pancreatic tumor.

My first decision was that I needed to take some specific dietary supplements and be on a controlled diet. As to the exact plan, I started to gather related information from several sources including the following:

a. literature search from everywhere possible,
b. my previous knowledge and experience, and
c. my research work with other collaborators.

After thorough research, I decided the following dietary supplements would be beneficial and that I would take them for at least one year until my next MRI.

My Dietary Supplements

Burdock Root and Goji Berries

The first dietary supplement I thought about was Burdock root, also called Gobo root (Fig. 4.1). I remembered reading about burdock root a few years ago saying that it might have anticancer effect. I therefore thought about taking advantage of it.

The long burdock root is a common Chinese/Japanese food ingredient. Although not commonly available in most grocery stores in the U.S., burdock root is available in most Chinese grocery stores.

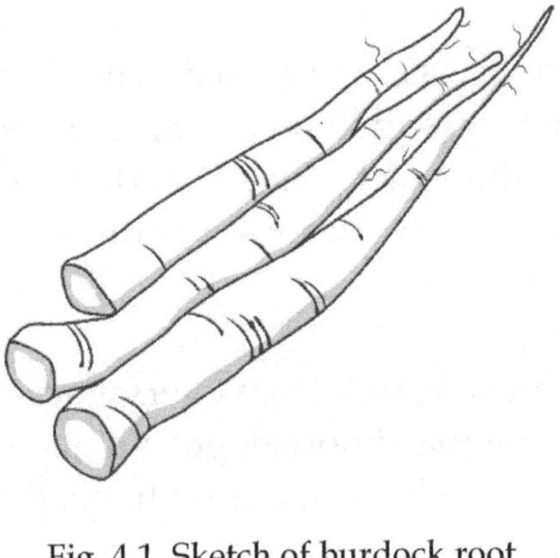

Fig. 4.1 Sketch of burdock root

It is believed that burdock root kills germs and also increases urine flow, therefore, it has been used to treat cold and high blood pressure. In addition, burdock root has also been used as a folk medicine for diabetes, joint pain, and cancer [10].

Burdock root contains arctiin, luteolin and quercetin [11]. Arctiin (Fig. 4.2) is a polyphenolic compound found in plants. It is reported to have shown anticancer effect *in vitro* (studies performed outside of a living organism) and in animal models [12]. Luteolin (Fig. 4.2) is a flavonoid that is also reported to be anticancer *in vitro* and could be useful in cancer prevention [13]. Quercetin is a flavonoid which will be discussed later.

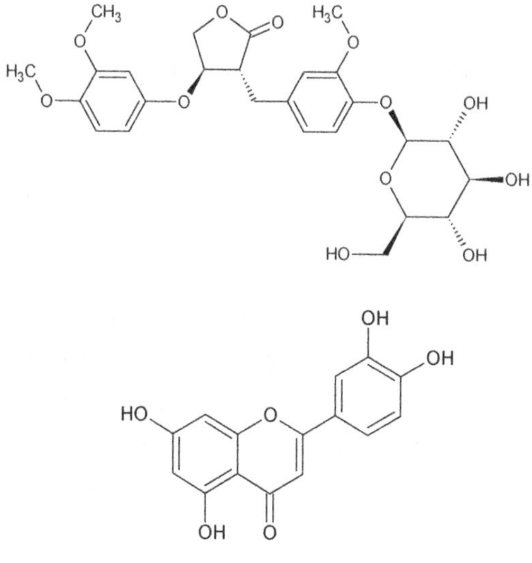

Fig. 4.2 Structures of arctiin (top) and luteolin (bottom)

Like many other natural products, burdock root has been made into different forms of dietary supplements, which are readily available commercially. Since I was able to purchase fresh burdock root, I did not buy the supplements comprising burdock root.

You can find a number of burdock root recipes on the internet. However, taking burdock root does not have to be complicated. For example, burdock root can be finely shredded and then added with some meat and desired spices to make a simple stir-fry with a pretty good taste.

My wife used a very simple recipe and made a special drink out of burdock root. Because burdock root is dug out of the ground, they tend to be kind of dirty. My wife would use a brush and water to thoroughly clean the outside of burdock root. After cleaning/washing burdock root, she would cut it into slices and place them in a slow cooker, often leaving them at a low heat overnight. It could then be ready as a healthful drink.

However, when I drank it, there was an earthy taste, which made the drink unattractive. We then decided to add a few goji berries in the drink. Because goji berries taste sweet, adding goji berries made the drink taste much better. I started to drink this special drink daily.

Adding goji berries not only made the drink much more drinkable, but also made the drink more nutritious. Goji berries are highly antioxidant and known to be anti-inflammatory due to its contents of quercetin, rutin, and other phenolic compounds [14]. Further, goji berries have been used in China for more than a thousand years and are generally pretty safe if taken in moderation. Please beware, goji berries may interact with certain drugs, including blood thinner drugs and diabetic drugs.

Here is one side effect of this special drink discovered by me. This special drink increases flatulence within a few hours; it could be because burdock root helps simulate the digestive system. Therefore, it is a good idea to drink it at night time.

Berberine

The second dietary supplement I took was berberine (Fig. 4.3). Berberine is found in the roots of a number of flowering plants called *Coptis* (Goadthread), such as coptis chinesis and coptis japonica. The species of *Coptis* in China is often referred to as HuangLian, thus, berberine is also called HuangLian Su (meaning extract from HuangLian).

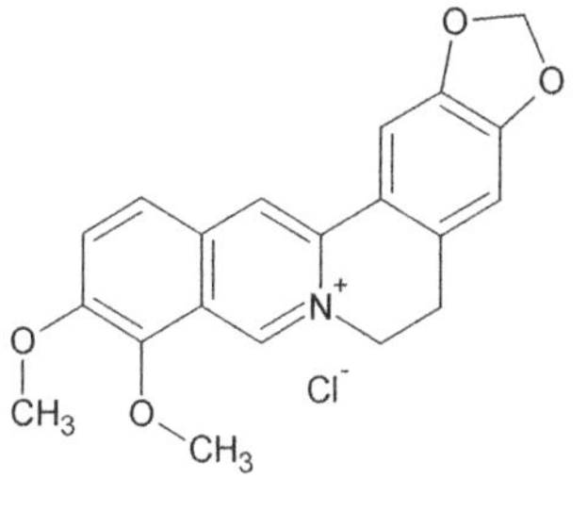

Fig. 4.3 Structure of berberine

HuangLian has been used for thousands of years as a traditional Chinese medicine for many indications, including bacterial & viral infections, diabetes, inflammation, liver injury, elevated blood cholesterol, and congestive heart failure. Beware that overdose of HuangLian may cause some side effects such as allergic reactions.

Berberine shows inhibitory effects on the proliferation and reproduction of certain tumorigenic microorganisms and viruses, such as *Helicobacter pylori* and hepatitis B virus [15].

In recent years, berberine has been reported to have anticancer activities both *in vitro* and *in vivo* (studies performed in a living organism) [15, 16]. Cell and animal studies indicated that it could have certain activities against a number of cancers including liver cancer, breast cancer, lung cancer, pancreatic cancer, prostate cancer, and melanoma [17-19].

In addition, the *in vitro* studies in the lab of Dr. Valeriote also indicated that berberine showed anticancer effects against several cancer cell lines.

Since berberine supplements were readily available from the market, I simply purchased berberine capsules from the market. I took two capsules, 400 mg/cap, two times a day (after breakfast and after dinner).

Vitamin D

I also decided to take vitamin D aggressively for a couple of reasons:

In 2007, I started to feel some pain when I slept on hard surfaces. This was kind of strange to me because I used to be sleeping on hard surface when I was young and never had any pain in my bones. Things got even tougher that I have to sleep on a very soft mattress; otherwise, I could not sleep because of the pain in my bones. Worse than that, when I was sitting too long, especially during a long flight, I could barely sit still. Because of the pain, I had to keep moving and shifting my hip and leg. It was really an awful experience.

After doing some research, I suspected that my symptom might be due to osteomalacia (softening of the bones) and I could be vitamin D deficient. In addition, vitamin D supplements could increase bone mineral density and be good for bone health [20]. Indeed, not too long after I started to take some high dose vitamin D regularly, the bone pain disappeared.

The second reason I continued to take vitamin D to combat my IPMN was based on some scientific reports. It was reported that vitamin D may have something to do with pancreatic IPMN. Recent scientific studies indicated that vitamin D and pancreatic cancer are closely related. Higher intakes of vitamin D are associated with lower risks for pancreatic cancer [21, 22]. In addition, there are also reports relating pancreatic cancer and vitamin D deficiency [23].

Based on the above two reasons, I suspected that my IPMN might be related to my long-term vitamin D deficiency. Although there is no concrete evidence to verify my suspicion, I decided it might be a good idea that I started to fortify my vitamin D intake.

Name: Jiajiu Shaw | DOB: [redacted] | MRN: [redacted] | PCP: [redacted]

VITAMIN D 25 HYDROXY (25HD) - Past Results

Name Standard Range	11/23/15	1/9/16
25-Hydroxy Vit D 25 - 100 ng/mL	58	>150

Fig. 4.4 My vitamin D levels on two different days

Almost immediately, I started to take vitamin D at a dose of 2x1,000 IU/day regularly. In the meantime, I asked my physician to test my vitamin D levels. One month after my IPMN was detected (11/23/2015), the blood test showed that my vitamin D level was at 58 ng/mL, within the normal range (25-100 ng/mL). Because I wanted to make sure that I have plenty of vitamin D, I started to take a much higher dose of vitamin D (2x10,000 IU/day) and another blood test on 1/9/2016 showed that my blood level of vitamin D was >150 ng/mL, which was over the normal limit (Fig. 4.4). My physician didn't like it and told me to reduce my intake of vitamin D. Since then, I did reduce the dose of vitamin D until November, 2016, but never stopped taking it. Even today, I am still taking some vitamin D from time to time but only at a much lower dose.

One interesting happening was that a couple months after I started to take the high dose vitamin D, my abdominal pain appeared to have disappeared. Was it due to vitamin D or the combination of both vitamin D and other supplements? Scientifically, it could also be due to other reason(s) and might not be due to what I was taking. No matter what, I was really happy that my abdominal pain was not there anymore.

Garlic

Garlic (*Allium sativum*) is a popular spice for most people. In addition, garlic has been reported to have a number of medicinal effects. For example, it has been used to help alleviate hypertension, high cholesterol, bronchitis, rheumatism, and diabetes.

Some clinical trials indicated that garlic may have anticancer effect [24]. Other studies indicate that increased intake of garlic is associated with reduced risk of certain cancer, including cancers of stomach, colon, esophagus, pancreas, and breast [25]. Clinical trials on the effect of allitridum (garlic extract) and selenium showed that risk of stomach cancer was significantly reduced [24].

Although I have not found a scientific report linking garlic and pancreatic IPMN directly, I thought it was probably not a bad idea to start taking garlic pills regularly.

Ginger

Ginger has been known to show anti-inflammatory, antioxidant and antiproliferative activities. In addition, it was shown that ginger may be used as a chemopreventive agent. A scientific report showed that whole ginger extract is effective against prostate cancer both *in vitro* and *in vivo* [26].

Further, a recent report showed that ginger extract has a potent anticancer activity against pancreatic cancer cells mainly through Reactive Oxygen Species-mediated cell death [27].

Ganoderma (LingZhi or Reishi Mushroom)

Ganoderma (Fig. 4.5) is a genus of polypore mushrooms which are wood-decaying fungi. It is one of the most popular dietary supplements in China to promote well-being and longevity and has been used as a traditional Chinese medicine for treating many diseases. In Chinese, LingZhi means "spiritual flower" although it is not really a flower. In fact, it is a flower-looking fungus with woody texture, which extracts nutrients from its hosting plant.

Ganoderma species have been investigated for a number of potential therapeutic benefits including antioxidant and anticancer effects, acting supposedly by strengthening the immune system and having liver-protecting effects [28-30]. Commercial products as dietary supplements in various forms, such as raw material or extract, are widely available.

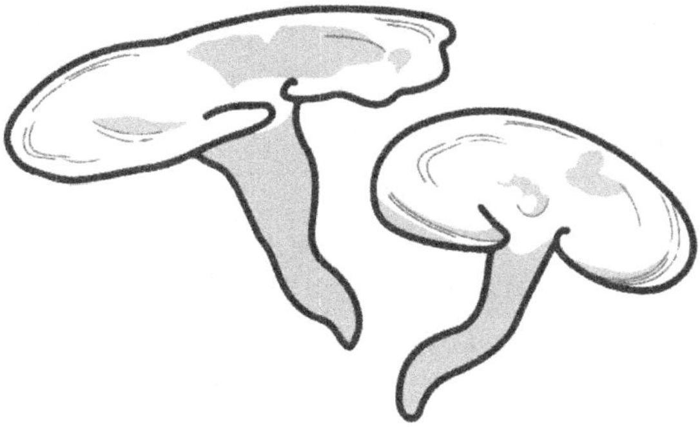

Fig. 4.5 *Ganoderma*

Quercetin

Quercetin (Fig. 4.6) is a flavonoid found in many plants including citrus fruit, tea, onion, and apples. It has been shown to have several biologically beneficial effects, including antioxidant, anti-inflammatory, and anticancer effects.

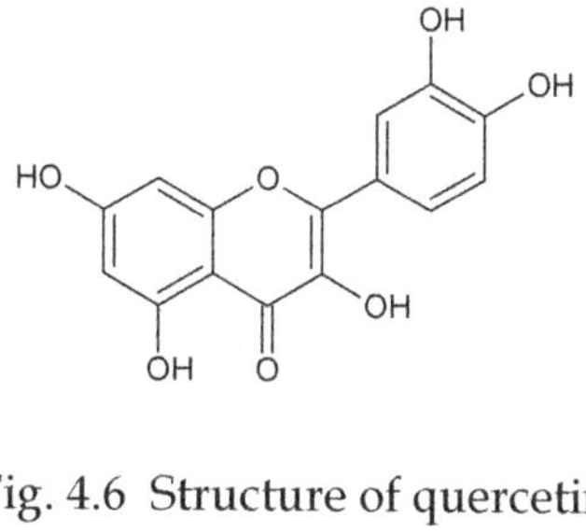

Fig. 4.6 Structure of quercetin

Most of the studies on quercetin and cancer have been done in cell culture (*in vitro* studies) or animal studies. *In vitro* studies indicate that quercetin can slow the growth of some cancer cells [31, 32]. One must be cautious that *in vitro* studies can be very different from animal studies in their results.

Clinical studies did indicate quercetin is involved in tyrosine kinase inhibition indicating that quercetin might be useful in cancer therapy [33, 34]. However, there is no solid clinical evidence indicating that quercetin can prevent or treat cancer.

<u>Chinese Skullcap (HuangQin in Chinese)</u>

Chinese skullcap is a plant native to China; its Chinese name is HuangQin. HuangQin has been used as a Chinese traditional medicine for thousands of years and reported to have certain beneficial biological activities against many diseases including cancer, asthma, allergy, arthritis, Alzheimer, and hepatoma.

There are more than 30 flavonoids in the root of Chinese skullcap; it is reported that baicalein flavonoid is one of the main active ingredients. Baicalein is one of the most studied herbal chemicals. It has been reported that both *in vitro* and *in vivo* studies indicated that baicalein showed potent activity against pancreatic cancer [35].

Genistein

Genistein (Fig. 4.7) is an isoflavone first isolated from *Genista tinctoria* (dyer's broom). It is also found in soy beans and soy-based products. It is widely available as a diet supplement; however, its use for therapeutic treatment is still under investigation.

Genistein was shown to have anticancer effects on various cancers including small cell lung cancer cells [36-38]. Genistein works through several mechanisms of action, including the inhibition of tyrosine kinases [39], which are important mediators of the signaling cascade.

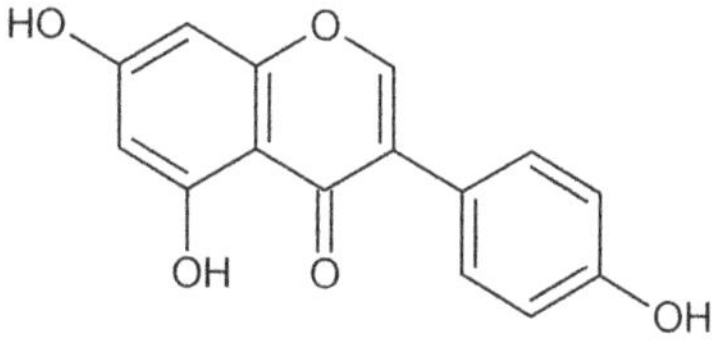

Fig. 4.7 Structure of genistein

Based on the reported properties of genistein, I did try to take genistein pills for a couple of weeks However, maybe I was taking too high a dose, I felt some abdominal pain, probably liver pain. The same thing happened when I took Chinese skullcap. In any case, I eventually decided to stop taking both genistein and Chinese skullcap.

Additional Notes

Since I am not here to promote any particular brand of dietary supplements, I won't provide the name of any manufacturers of dietary supplements. As you may know, natural product supplements are generally not as highly regulated as drugs. Therefore, the same supplement made by different manufacturers might be different in the purity and/or content. Just like most other consumers, I had to trust the label provided by the manufactures.

Furthermore, one must be careful in distinguishing the difference between the raw natural products and their extracts. For example, some supplements are made simply by grinding the leaves; some are made by soaking natural products in a solvent (e.g., water or ethanol) at a certain temperature for a certain period of time so that only certain ingredients are extracted and then dried up (thus often referred to as extracts). Again, different companies could be using different extraction methods, thus, an extract of one natural product made by different companies could also be somewhat different in their contents.

The following list is a brief summary of what I was taking every weekday during the one year period.

After breakfast and after dinner

Vitamin D: 2 x 1,000 IU capsule (11/1 - 11/23/2015); 2 x 10,000 IU capsule (11/24/2015 - 1/9/2016); 1 x 1,000 IU capsule (from 1/10/2016)

Berberine: 2 x 400 mg capsules after breakfast and after dinner (from 11/1/2015)

Garlic: 2 x 1,000 mg softgels after breakfast and after dinner (from 12/15/15)

Ginger: 2 x 550 mg capsules after breakfast and after dinner (from 12/15/15)

Between breakfast and lunch

Quercetin: 2 x 800 mg capsules

Nattokinase: 2 x 100 mg capsules (2,000 FU/cap)

During Dinner

Burdock/goji drink: one cup (about 300 mL)

After dinner

Vitamin D: 2 x 1,000 IU capsule (11/1 - 11/23/2015); 1 x 10,000 IU capsule (11/24/2015 - 1/9/2016); 1 x 1,000 IU capsule (from 1/10/2016)

Berberine: 2 x 400 mg capsules after breakfast and after dinner (from 11/1/2015)

Garlic: 2 x 1,000 mg softgels after breakfast and after dinner (from 12/15/15)

Ginger: 2 x 550 mg capsules after breakfast and after dinner (from 12/15/15)

Looking at the list, most people would say it's extreme. So did I, but I felt desperate. Nevertheless, it was pretty much what I was taking for about one year. I did try to give my digestive system a break during the weekend and limited my intake of these supplements.

The reason that quercetin and nattokinase were in my list was not directly related to my IPMN. I was already taking quercetin regularly for general maintenance of my health, well before this ordeal.

Why was I taking nattokinase? It was because I used to have high cholesterol (220 - 250 mg/dL); nattokinase was reported to be able to reduce blood clotting and fibrinogen, thus may be a potential nutraceutical for cardiovascular diseases [40]. With the help of Mr. Joseph Media, I also did a simple *in vitro* test which showed that nattokinase was effective in both preventing and dissolving blood clotting.

After taking nattokinse for some time, my blood cholesterol was significantly lowered. Based on the blood test done in January 2016, my blood cholesterol was down to 180 mg/dL, much lower than what I used to have.

Other Dietary Actions

In addition to taking the above supplements, I also attempted to control my other food intake. Some of the activities I took are described below.

As discussed in my previous book (IS CANCER IN US), too much sugar is not good for one's health especially for cancer patients. Although cutting down my intake of dessert and cookies was not easy, I had to do it. Therefore, one of the principles that I always wanted to follow was to reduce the consumption of sugar.

In addition to cutting down dessert and cookies, I also decided to stop drinking soft drinks. In the one year period, I probably drank a total of only two or three cans of soft drink. The reason was there was too much sugar in each can (probably 20 - 30 g of sugar in each can). Furthermore, I realized that in most bottled juices, there was also too much sugar. What I decided to do was to dilute the juice with some water (1:1) before I drank the juice although it became less tasty and I did not always remember to do that. Overall, I was trying to reduce my total sugar intake. As you may know, eating too much sweet food is not good for one's health, especially for some cancer patients.

Furthermore, I also tried to reduce the intake of fatty foods. It is known that high-fat dietary products are not good for the pancreas especially for a patient with pancreatitis (inflammation of the pancreas). Because my wife enjoyed cooking meaty food, we had to consciously reduce the amount of meat in cooking, basically by reducing the amount of meat to one-half when cooking. As a result, I lost more than 5 lb after the one year period. I was not sure if it's good or bad, but that was a fact.

The other factor that could affect pancreas health is alcohol. Excessive consumption of alcoholic beverages is not good for someone with pancreatic issues, such as pancreatitis. Since I rarely had any alcoholic drink, it was not too hard for me to keep away from alcoholic drinks.

Caution! I always believe that there is a limit to everything. If something is overdone, there will be undesirable consequences. The same principle works in taking dietary supplements. For example, nattokinase should be taken with caution, preferably after having discussions with medical professionals.

Chapter 5

What Happened Then

The Verdict

One year after discovering that I had an IPMN, I went for my second MRI test on November 4, 2016. One day later, I received an e-mail telling me that the test results were available.

With some trepidation, I hesitated before I slowly opened up the MRI results on the internet. In the beginning of the report, it was my physician's summary and he did not talk about the pancreas. As I continued to read the results with great curiosity, I found the description about my pancreas (Fig. 5.1), which says:

> *Spleen, pancreas and adrenal glands are unremarkable. No pancreatic ductal dilatation. Previously seen subcentimeter cystic lesion in the pancreatic neck is not visualized.*

Name: Jiajiu Shaw | DOB: [redacted] | MRN: [redacted] | PCP: [redacted]

MR LIVER - Details

Narrative

MRI abdomen with and without contrast dated 11/4/2016.

[redacted]

Comparison: Ultrasound dated 4/29/2016; MR dated 10/24/2015.

Procedure: MRI abdomen with and without contrast. Routine liver protocol. 20 mL MultiHance injected IV. No immediate complication. 1.5 Tesla magnet.

Findings:

[redacted]

There is a 0.4 cm gallbladder polyp (series 201, image 12) corresponding to comparison ultrasound finding. Spleen, pancreas and adrenal glands are unremarkable. No pancreatic ductal dilatation. Previously seen subcentimeter cystic lesion in the pancreatic neck is not visualized.

Fig. 5.1 Results of my MRI on 11/4, 2016

"Oh my God!" I was so relieved. My pancreatic tumor was not there anymore; it was gone. Although I had several cysts on both my liver and kidneys, but they didn't seem to be too serious and I was not as concerned at this time. The results were better than what I was hoping for; I was only hoping that my IPMN could be reduced in its size.

After reading my MRI results, I was pleasantly surprised. My pancreatic IPMN was not detectable, better than what I was hoping for. I believed that this was most likely due to my "crazy" plan.

As we all know, nothing is perfect in life. My "crazy" plan could not be all good. Indeed, it was the case.

Looking back, I suffered several side effects, which might be due to my action. The following side effects are likely due to my regimen.

Potential Side Effects

Abdominal Pain

I started to take Chinese skullcap (2 x 400 mg/cap, 2 times per day) from 12/11/15. In about a week or so, I felt abdominal pain in the upper section. Was it my liver? Could it be because I was taking too high a dose? I wasn't sure. As a result, I decided to stop taking it and the pain disappeared shortly thereafter.

Also, I started to take genistein sometime in January of 2016 and I felt similar pain in my upper abdominal area. Just like in the case for Chinese skullcap, I was not sure why. Was it because I was taking too many other supplements which caused some interactions with genistein or was it something else?

It has been reported that genistein protects acetaminophen-induced acute liver injury [41] and prevents oxidative liver damage [42]. On the other hand, genistein is a tyrosine kinase inhibitor and tyrosine kinase inhibitors are known to have hepatotoxicity (toxicity to the liver) [43]. Both reports were supported by scientific studies; it was not very clear as to what had happened to me.

In any case, after I stopped taking genistein, my upper abdominal pain disappeared in a short time.

Joint Pain

As most people know, just like medicinal drugs, all natural products have certain side effects especially when too much is taken. The difference is that the side effects could be different or to a lower extent. Therefore, it is important that one should be cautious in taking any supplements.

At some point in time during the one year period, I started to feel some pain in my joints, especially my knees and hips. It got worse when I walked fast or walk upstairs; I even decided to take an elevator to my lab on the 4th floor instead of taking the stairs, which was my previous daily ritual.

Even worse, in my recent trip to China (in the second half of October, 2016), I did a lot of walking during the two-week trip. Almost every day, my knees were hurting; sometimes, my hip also hurt. This was something I never experienced before.

In fact, I had no idea whether it was from my excessive walking or from my taking too many dietary supplements. Assuming it was from the supplements, I might have to significantly reduce my intake of supplements. In addition, I also had to take something else to tackle my knee issue.

Two supplements were on my list of selection to help improve the health of my knees and joints: glucosamine/chondroitin/MSM and calcium with vitamin D. Since I knew that my IPMN was gone, I started to take glucosamine/chondroitin/MSM and calcium supplement. As a result, my knees and hips seemed to be in a better shape now. Brief descriptions of both supplements are shown below.

Glucosamine/chondroitin/MSM

Glucosamine is a natural product found in both cartilage and synovial fluid. It has been approved in Europe for treating osteoarthritis and promoting joint health. In the U.S., glucosamine has been sold as a dietary supplement for years in different salt forms, including hydrochloride and sulfate. Most people believe it is effective in relieving joint pains. However, depending upon the formulation and how clinical trials are conducted, the symptomatic relief varies [44].

Chondroitin and MSM (methylsulfonylmethane) both have been used for osteoarthritis and commonly used together with glucosamine as a popular combination supplement to relieve joint pain.

Calcium

Almost everyone knows that calcium plays an important role in healthy bone. Because the body cannot produce calcium, it needs to be provided through the food we eat or supplements we take. If calcium levels are too low, our bone density can be low and the bones can become weak.

It has been reported that there is a synergistic effect between calcium and vitamin D. In order to decrease fracture incidence, clinical trials have shown that vitamin D and calcium together had better results than vitamin D or calcium alone [20, 45]. It is therefore believed that taking vitamin D together with calcium may be more effective in promoting bone health.

Urinating issue

About six months after I began my dietary plan, I started to have issues with my urinary system. When I had an urge to urinate, it was almost too late. I had to find a washroom almost right away. Also, I seemed to need to go to the washroom more often at night time. At first, I thought maybe it was simply because I was getting old. However, after a thorough recollection, it became clear that I wasn't in such a bad shape before I started my supplement plan.

After November of 2016, I essentially stopped my special dietary plan. In about 10 days or so, my urinating situation was much better already. It became clear that my urinating issue was probably due to my dietary plan; I suspected that it could be due to one or more of the supplements I was taking. Another possible reason is that maybe I was taking too many dietary supplements, which was too high a burden on my kidneys.

Fatigue

I used to be alert all day long. However, during the one year period that I was on my special dietary plan, I became tired more easily. Most every afternoon, I felt sleepy. There was a good chance that this could be due to my special dietary plan. I suspect that ginger might be one of the culprits, but I am not sure. Now that I am not on this plan anymore, I don't fatigue as easily.

Weight Loss

I never really paid too much attention to my weight, but I did remember that I weighed slightly less than 150 lb. During 2015 and 2016, I lost a little bit weight to around 140 lb. I did not know whether it was due to my pancreatic IPMN or due to my intake of too many dietary supplements. Nevertheless, I have reduced the amount of supplements.

Another Incident

In late November, 2016, I suffered severe abdominal pain and went to the emergency room for almost a whole day. After a series of testing, it was concluded that my abdominal pain was due to two factors: severe constipation and a kidney stone. During the process, a CT scan was also performed. Several organs were scanned including the pancreas. Again, the report said that my pancreas was unremarkable (Fig. 5.2). This confirmed my MRI results in early November of 2016, which indicted that my pancreas was unremarkable.

Document info

Result type:	CT Abdomen+Pelvis w IV Con
Result date:	Nov 28, 2016, 10:10 p.m.
Result status:	authenticated
Performed by:	[redacted]
Modified by:	[redacted]
Accession number:	[redacted]

CT Abdomen+Pelvis w IV Con

Patient: JIAJIU SHAW **DOB:** [redacted]

CT Abdomen+Pelvis w IV Con 11/28/2016 22:10

FINDINGS:

[redacted]

Pancreas: Unremarkable.

[redacted]

Fig. 5.2 Results of my CT scan on 11/28, 2016

Chapter 6

What Next

Future Plan

What should I be doing now? What is my plan for the future?

Recalling that I was shocked and in despair just over one year ago, I am now really grateful that my IPMN is not there anymore. However, it is probably not a good idea to be complacent; therefore, I should still take some actions to reduce the chance of having to deal with the same situation.

My plan for the near future can be briefly described below:

1. Most people probably would agree that my previous one-year action was very empirical. Nevertheless, I don't plan to totally give it up. My plan is to maintain the habit of taking some of the dietary supplements I was taking. For example, I will continue to take the special burdock root/goji drink at least five days a week. I will also take a couple of garlic pills (1,000 mg/cap) every other day.

2. I will continue to take vitamin D on a regular basis but not the extra high dose. I also plan to take multivitamin regularly. This is something I did not pay too much attention to in the past. However, at my age, it is probably a good idea to do that so that I won't be missing some key vitamins.

3. I will continue to reduce the consumption of sweet dessert, but I may indulge my appetite for sweets sometimes.

4. I also plan to reduce the consumption of junk food, especially high fat and deep-fried foods. This is because I believe this action can reduce the burden on my pancreas and may be good for my general health.

5. It makes sense to do regular exercise for personal health. My plan is to do regular daily exercise for about 20 minutes a day, including walking up to the 4th floor to my office, regular stretching during the work hours, and some Tai Chi.

Final Notes

I believe that we can all learn some lessons from some incidents in our lives. With this special experience, I believe I have learned the following lessons with respect to personal health:

First of all, one must have a plan in order to improve one's health. The plan should include at least proper diet and regular exercise. I cannot emphasis enough the importance of both proper diet and suitable exercise. In addition, I also believe that mental status is very important.

As you may know, all biological activities are controlled and/or related to many signal transduction pathways directly or indirectly. A person's mental status can enhance or inhibit certain signal pathways, thus, one's health. Therefore, positive mental status is very important for a healthy body.

Secondly, a combination of certain supplements or specific supplements may be helpful in treating certain tumors. Maybe I am biased, but I believe a combination of several active components should be better than one single component. This is because the human body is such a complex biological system that is controlled by numerous tangible biological components and intangible signal transduction pathways; they are mostly intertwined in a delicate and complicated manner. Can one single agent achieve its beneficial physiological effect by modulating one single biological signal pathway? Chances are low. A suitable combination should work better than a single agent. Maybe some of the supplements could have been removed from my long list and still achieve the same effect (removal of my IPMN)? Maybe, but I don't really know the answer.

The third lesson I learned is that we must be persistent in executing our plans. No matter how good a plan is, if one does not take actions and follow through on his plan, the plan is useless. After all, if we do not take action according to our plans, there won't be any good results.

The fourth lesson is that, dietary supplements, if taken in excess, can also produce significant side effects. Even for FDA approved drugs, all of them are associated with certain toxicities if not taken according to the instructions. To achieve the desired therapeutic effect, a drug has to be taken at a proper dose according to a suitable regimen. If taken properly, the efficacy can be maximized and the side effect can be minimized. This is the same situation for dietary supplements. If not properly taken, all dietary supplements could cause undesirable side effects.

Last but not least, in order to deal with one's health situation, a patient must work closely with his/her physician.

Finally, the old Chinese saying, **Do one's best and leave the rest to God**, makes a lot of sense to me. To deal with one's health, one must do one's best in all manners possible and leave the rest to the all mighty Nature since there are things that we can't control.

References:

1. Terris, B., et al., *Intraductal papillary mucinous tumors of the pancreas confined to secondary ducts show less aggressive pathologic features as compared with those involving the main pancreatic duct.* Am J Surg Pathol, 2000. **24**(10): p. 1372-7.

2. Xiao, S.Y., *Intraductal papillary mucinous neoplasm of the pancreas: an update.* Scientifica (Cairo), 2012. **2012**: p. 893632.

3. Khan, S., G. Sclabas, and K.M. Reid-Lombardo, *Population-based epidemiology, risk factors and screening of intraductal papillary mucinous neoplasm patients.* World J Gastrointest Surg, 2010. **2**(10): p. 314-8.

4. Grutzmann, R., et al., *Intraductal papillary mucinous tumors of the pancreas: biology, diagnosis, and treatment.* Oncologist, 2010. **15**(12): p. 1294-309.

5. Sohn, T.A., et al., *Intraductal papillary mucinous neoplasms of the pancreas: an updated experience.* Ann Surg, 2004. **239**(6): p. 788-97; discussion 797-9.

6. Jang, J.Y., et al., *Treatment guidelines for branch duct type intraductal papillary mucinous neoplasms of the pancreas: when can we operate or observe?* Ann Surg Oncol, 2008. **15**(1): p. 199-205.

7. *Laparoscopic Pancreas Surgery at Johns Hopkins.* 2016 [cited 2016 Dec 10, 2016]; Available from: http://pathology.jhu.edu/pancreas/TreatmentLaparoscopic.php.

8. Topazian, M., et al., *Photodynamic therapy of intraductal papillary mucinous neoplasm.* Endoscopy, 2012. **44**(2): p. 213-5.

9. Ikeuchi, N., et al., *Prognosis of cancer with branch duct type IPMN of the pancreas.* World J Gastroenterol, 2010. **16**(15): p. 1890-5.

10. Chan, Y.S., et al., *A review of the pharmacological effects of Arctium lappa (burdock).* Inflammopharmacology, 2011. **19**(5): p. 245-54.

11. Ferracane, R., et al., *Metabolic profile of the bioactive compounds of burdock (Arctium lappa) seeds, roots and leaves.* J Pharm Biomed Anal, 2010. **51**(2): p. 399-404.

12. Matsuzaki, Y., et al., *Arctiin induces cell growth inhibition through the down-regulation of cyclin D1 expression.* Oncol Rep, 2008. **19**(3): p. 721-7.

13. Lin, Y., et al., *Luteolin, a flavonoid with potential for cancer prevention and therapy.* Curr Cancer Drug Targets, 2008. **8**(7): p. 634-46.

14. Nardi, G.M., et al., *Anti-inflammatory Activity of Berry Fruits in Mice Model of Inflammation is Based on Oxidative Stress Modulation.* Pharmacognosy Res, 2016. **8**(Suppl 1): p. S42-9.

15. Sun, Y., et al., *A systematic review of the anticancer properties of berberine, a natural product from Chinese herbs.* Anticancer Drugs, 2009. **20**(9): p. 757-69.

16. Diogo, C.V., et al., *Berberine as a promising safe anti-cancer agent - is there a role for mitochondria?* Curr Drug Targets, 2011. **12**(6): p. 850-9.

17. Tang, J., et al., *Berberine and Coptidis rhizoma as novel antineoplastic agents: a review of traditional use and biomedical investigations.* J Ethnopharmacol, 2009. **126**(1): p. 5-17.

18. Serafim, T.L., et al., *Different concentrations of berberine result in distinct cellular localization patterns and cell cycle effects*

in a melanoma cell line. Cancer Chemother Pharmacol, 2008. **61**(6): p. 1007-18.

19. Pandey, M.K., et al., *Berberine modifies cysteine 179 of IkappaBalpha kinase, suppresses nuclear factor-kappaB-regulated antiapoptotic gene products, and potentiates apoptosis.* Cancer Res, 2008. **68**(13): p. 5370-9.

20. Lips, P. and N.M. van Schoor, *The effect of vitamin D on bone and osteoporosis.* Best Pract Res Clin Endocrinol Metab, 2011. **25**(4): p. 585-91.

21. Skinner, H.G., et al., *Vitamin D intake and the risk for pancreatic cancer in two cohort studies.* Cancer Epidemiol Biomarkers Prev, 2006. **15**(9): p. 1688-95.

22. Barreto, S.G. and R.E. Neale, *Vitamin D and pancreatic cancer.* Cancer Lett, 2015. **368**(1): p. 1-6.

23. Garland, C.F., et al., *Cloud cover-adjusted ultraviolet B irradiance and pancreatic cancer incidence in 172 countries.* J Steroid Biochem Mol Biol, 2016. **155**(Pt B): p. 257-63.

24. Li, H., et al., *An intervention study to prevent gastric cancer by micro-selenium and large dose of allitridum.* Chin Med J (Engl), 2004. **117**(8): p. 1155-60.

25. Fleischauer, A.T. and L. Arab, *Garlic and cancer: a critical review of the epidemiologic literature.* J Nutr, 2001. **131**(3s): p. 1032S-40S.

26. Karna, P., et al., *Benefits of whole ginger extract in prostate cancer.* Br J Nutr, 2012. **107**(4): p. 473-84.

27. Akimoto, M., et al., *Anticancer Effect of Ginger Extract against Pancreatic Cancer Cells Mainly through Reactive Oxygen Species-Mediated Autotic Cell Death.* PLoS One, 2015. **10**(5): p. e0126605.

28. Yuen, J.W. and M.D. Gohel, *Anticancer effects of Ganoderma lucidum: a review of scientific evidence.* Nutr Cancer, 2005. **53**(1): p. 11-7.

29. Xu, Z., et al., *Ganoderma lucidum polysaccharides: immunomodulation and potential anti-tumor activities.* Am J Chin Med, 2011. **39**(1): p. 15-27.

30. Sanodiya, B.S., et al., *Ganoderma lucidum: a potent pharmacological macrofungus.* Curr Pharm Biotechnol, 2009. **10**(8): p. 717-42.

31. Deng, X.H., et al., *Effects of quercetin on the proliferation of breast cancer cells and expression of survivin.* Exp Ther Med, 2013. **6**(5): p. 1155-1158.

32. Gulati, N., et al., *The antiproliferative effect of Quercetin in cancer cells is mediated via inhibition of the PI3K-Akt/PKB pathway.* Anticancer Res, 2006. **26**(2A): p. 1177-81.

33. Ferry, D.R., et al., *Phase I clinical trial of the flavonoid quercetin: pharmacokinetics and evidence for in vivo tyrosine kinase inhibition.* Clin Cancer Res, 1996. **2**(4): p. 659-68.

34. Arora, A. and E.M. Scholar, *Role of tyrosine kinase inhibitors in cancer therapy.* J Pharmacol Exp Ther, 2005. **315**(3): p. 971-9.

35. Donald, G., K. Hertzer, and G. Eibl, *Baicalein--an intriguing therapeutic phytochemical in pancreatic cancer.* Curr Drug Targets, 2012. **13**(14): p. 1772-6.

36. Tian, T., et al., *Genistein exhibits anti-cancer effects via down-regulating FoxM1 in H446 small-cell lung cancer cells.* Tumour Biol, 2014. **35**(5): p. 4137-45.

37. Li, Q.S., et al., *Genistein and its synthetic analogs as anticancer agents.* Anticancer Agents Med Chem, 2012. **12**(3): p. 271-81.

38. Li, H.Q., Y. Luo, and C.H. Qiao, *The mechanisms of anticancer agents by genistein and synthetic derivatives of isoflavone.* Mini Rev Med Chem, 2012. **12**(4): p. 350-62.

39. Nakashima, S., T. Koike, and Y. Nozawa, *Genistein, a protein tyrosine kinase inhibitor, inhibits thromboxane A2-mediated human platelet responses.* Mol Pharmacol, 1991. **39**(4): p. 475-80.

40. Hsia, C.H., et al., *Nattokinase decreases plasma levels of fibrinogen, factor VII, and factor VIII in human subjects.* Nutr Res, 2009. **29**(3): p. 190-6.

41. Fan, Y.J., et al., *Genistein protection against acetaminophen-induced liver injury via its potential impact on the activation of UDP-glucuronosyltransferase and antioxidant enzymes.* Food Chem Toxicol, 2013. **55**: p. 172-81.

42. Mohamed Salih, S., et al., *Genistein improves liver function and attenuates non-alcoholic fatty liver disease in a rat model of insulin resistance.* J Diabetes, 2009. **1**(4): p. 278-87.

43. Shah, R.R., J. Morganroth, and D.R. Shah, Hepatotoxicity of tyrosine kinase inhibitors: clinical and regulatory perspectives. Drug Saf, 2013. 36(7): p. 491-503.

44. Henrotin, Y., A. Mobasheri, and M. Marty, Is there any scientific evidence for the use of glucosamine in the management of human osteoarthritis? Arthritis Res Ther, 2012. 14(1): p. 201.

Made in the USA
Monee, IL
04 November 2026

67386989R00036